FORCES

By
Robin Twiddy

BookLife PUBLISHING

Written by:
Robin Twiddy

Edited by:
Madeline Tyler

Designed by:
Amy Li

©2023
BookLife Publishing Ltd.
King's Lynn, Norfolk
PE30 4LS, UK

A catalogue record for this book is available from the British Library.

ISBN: 978-1-80155-824-2
ISBN: 978-1-80155-855-6

All facts, statistics, web addresses and URLs in this book were verified as valid and accurate at time of writing.
No responsibility for any changes to external websites or references can be accepted by either the author or publisher.

IMAGE CREDITS

All images are courtesy of Shutterstock.com, unless otherwise specified. With thanks to Getty Images, Thinkstock Photo and iStockphoto.

Recurring images – VectoRay, Easy Vector art idea (patterns). Cover – Africa Studio, Fabrika Simf, revers, Engineer studio, Peter Vanco, Ballz3389. 1 – Fabrika Simf. 2–3 – Africa Studio, Fabrika Simf, revers, Engineer studio, Peter Vanco. 4–5 – Helen Nertis, Ppstock, Secheltgirl, Nikolaevap. 6–7 – NotionPic, Akkalak Aiempradit, Irina Wilhauk. 8–9 – Amorn Suriyan, Colorfuel Studio, FamVeld, NotionPic, ONYXprj. 10–11 – BNP Design, Jacob Lund, Robert Kneschke. 12–13 – ClassicVector, NotionPic, Sergey Novikov, wavebreakmedia. 14–15 – Helen Nertis, BNP Design Studio, LeManna, TinnaPong. 16–17 – Denys Drozd, Elizaveta Galitckaia, graphic-line, Nadya Eugene. 18–19 – Good Studio, FARBAI, Paul Aniszewski, SkyLynx. 20–21 – 777 Bond vector, Martin Novak, Rawpixel.com. 22–23 – Olesia Bilkei, Petite usagi, tanyabosyk. 24 – Peter Vanco, revers.

CONTENTS

PAGE 4	What Are Forces?
PAGE 6	Push
PAGE 8	Jump
PAGE 10	Pull
PAGE 12	Climbing
PAGE 14	Speed
PAGE 16	Scooters
PAGE 18	Invisible Forces
PAGE 20	Gravity
PAGE 22	Forces All Around
PAGE 24	Glossary and Index

Words that look like <u>this</u> can be found in the glossary on page 24.

WHAT ARE FORCES?

Forces make things move. There are many different forces. We use lots of them every day.

Two types of force we use all the time are pushing and pulling forces.

This child is using a pulling force to move this dog in a wagon.

PUSH

When we push, we use our bodies to move something away from us.

Push

We can push things harder or softer to make them move faster or slower.

JUMP

When you jump, you use your legs to push your body away from the ground.

How high can you jump?

If you use more force to push, you will jump higher.

PULL

When we pull, we are using a pulling force to move something closer to us.

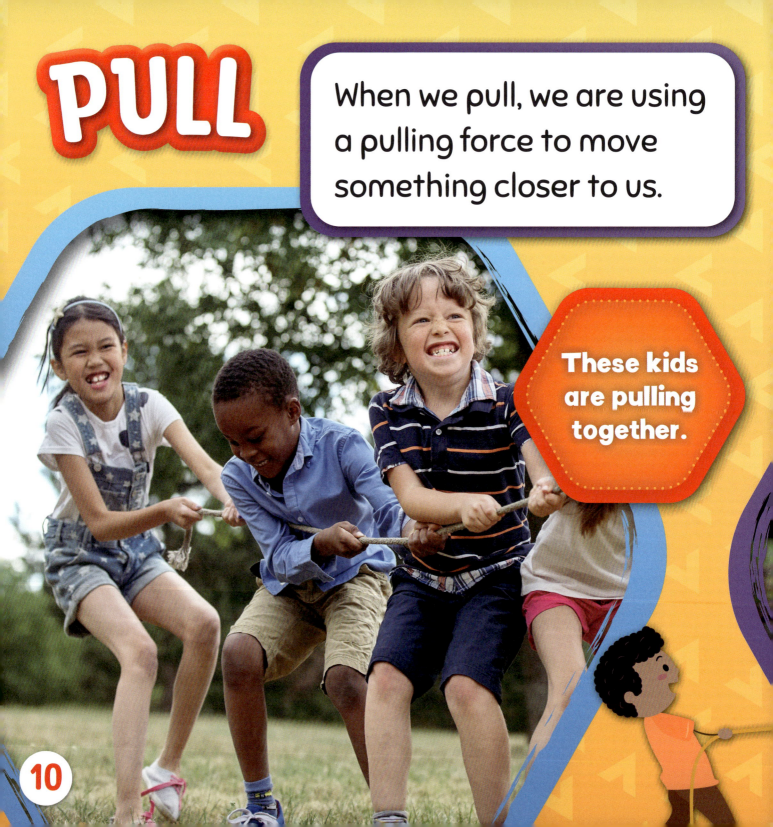

These kids are pulling together.

When we pull, we can change the <u>speed</u> and <u>direction</u> of something.

Pulling harder or softer will change the speed of the thing you are pulling.

Pull

CLIMBING

Have you ever tried to climb a rope? You need to pull hard to make your body go up.

Pulling down

This boy is pulling with his arms and pushing with his legs to climb the climbing wall.

SPEED

You can use force to make something speed up or slow down.

Pushing on the peddles of a bike will make it speed up.

Force can be used to turn a bike by pushing and pulling on the handlebars.

Turning the handlebars turns the front wheel of the bike.

SCOOTERS

Do you know what forces you are using when you ride on a scooter?

You push on the ground to move the scooter forward.

INVISIBLE FORCES

We cannot always see the things that are making the forces around us.

The wind is making a pushing force that moves the trees.

Wind can push with a very strong force. Wind turbines are moved around by the pushing force of the wind.

GRAVITY

Gravity is a force that you cannot see, but you can see how it works all the time.

Gravity is a force that pulls down.

When you jump up on a trampoline, gravity pulls you back down.

Gravity

Gravity is pulling everything on <u>Earth</u> down all the time.

FORCES ALL AROUND

There are forces all around us all the time and we use forces in everything we do.

A pushing force has changed the shape of this dough.

There are many more forces to learn about.

What kinds of forces have you used today?

GLOSSARY

DIRECTION the way that someone or something is moving, for example, right, left, up or down

EARTH the planet that we live on

SPEED how fast or slow something is moving

INDEX

BIKES 14–15

DIRECTIONS 11

DOUGH 22

PULLS 5, 10–13, 15, 20–21

PUSHES 5–9, 13–15, 17–19, 22

SLOW 7, 14

SPEED 11, 14

WIND 18–19